WEIRD SCIENCE

Bizarre Experiments and Strange Discoveries

Stuart A. Kallen

ReferencePoint Press

San Diego, CA

For more information, contact:
ReferencePoint Press, Inc.
PO Box 27779
San Diego, CA 92198
www.ReferencePointPress.com

LIBRARY OF CONGRESS CATALOGING-IN-PUBLICATION DATA

Names: Kallen, Stuart A., 1955- author
Title: Weird science : bizarre experiments and strange discoveries / by Stuart A. Kallen.
Description: San Diego, CA : ReferencePoint Press, Inc, [2027] | Includes bibliographical references and index. | Audience term: Teenagers | Audience term: High school students
Identifiers: LCCN 2026000579 (print) | LCCN 2026000580 (ebook) | ISBN 9781678213824 library binding | ISBN 9781678213831 ebook
Subjects: LCSH: Science--Experiments | Science--Research | Discoveries in science
Classification: LCC Q163 .K335 2027 (print) | LCC Q163 (ebook) | DDC 502.8--dc23/eng/20260407
LC record available at https://lccn.loc.gov/2026000579
LC ebook record available at https://lccn.loc.gov/2026000580

CONTENTS

The Ig Nobel Prizes have been poking gentle humor at the respected Nobel Prizes since 1991. Nobel Prizes are handed out to the world's most respected physicists, chemists, doctors, and other researchers. The Ig Nobels (a pun on *ignoble*, meaning "dishonorable") celebrate strange or unusual research. A sampling of prizes proves that scientific research can be weird, funny, and intellectually challenging.

1991: The first Ig Nobel Prize in Chemistry was awarded to a writer who theorized that water has an intelligence that allows it to remember substances it touched and events that have long passed.

2000: The physics award was granted to biologists who used magnets to levitate frogs.

2001: A doctor won the medicine prize for his report on physical injuries caused by coconuts falling from trees.

2004: The Japanese inventor of karaoke was awarded the peace prize after showing that singing karaoke provides a unique method for people to learn to endure one another.

2005: The nutrition prize was won by a Tokyo doctor who photographed every meal he consumed for thirty-four years.

2007: The literature prize went to an Australian linguist who studied how the word *the* causes problems for people trying to put things in alphabetical order.

2009: Two English researchers won the veterinary medicine prize for proving that cows with names produce more milk than unnamed cows.

2012: A Dutch researcher won the anatomy prize after proving that chimpanzees can identify other chimpanzees by looking at photographs of their rear ends.

2013: The Ig Nobel Prize in Medicine was given to Japanese scientists who analyzed the healing effects of opera music on mice who had heart transplants.

2014: The prize in Arctic science went to researchers who studied how reindeer react when approached by humans wearing polar bear costumes.

2016: German researchers were given the medicine prize by proving a person who has an itch on the left side of the body can find relief by looking into the mirror and scratching the right side of the body.

2017: A British researcher won the anatomy prize for medical research that produced the paper "Why Do Old Men Have Big Ears?"

2018: The peace prize was given to Spanish researchers who measured the psychological benefits of yelling and swearing at others while driving a car.

2019: Portuguese scientists who measured how well human spit cleans dirty surfaces were awarded the chemistry prize.

2020: Dutch researchers received the medicine prize for recognizing an undiagnosed condition called misophonia, which causes suffering in those who hear others chewing.

2021: A group of researchers were awarded the kinetics prize after studying why pedestrians sometimes collide with one another.

2022: Swedish scientists won the safety engineering prize for developing a moose crash dummy used to test ways to reduce damage from car–moose collisions.

2023: A Polish paleontologist was awarded two Ig Nobel Prizes, one in chemistry and the other in geology, for his paper that explained why scientists like to lick rocks.

2024: A team of Japanese researchers proved that some mammals are capable of breathing through their anuses, a breakthrough that won them the Ig Nobel Prize in Physiology.

2025: William B. Bean was awarded the Ig Nobel Prize in Literature for his journal, which described and analyzed the growth of one of his fingernails over a period of thirty-five years.

Strange and Surprising Experiments

Most people are familiar with the Nobel Prizes. These prestigious awards are granted annually to doctors, scientists, and others who have made transformative discoveries in the fields of chemistry, physics, or medicine. Nobel Prize winners are viewed as brilliant researchers at the peak of their profession. Their names appear in history books alongside renowned scientists like physicist Albert Einstein and medical researcher Jonas Salk.

Fewer people have heard of the Ig Nobel Prizes, created by a company called Improbable Research. According to the company, Ig Nobel Prizes are awarded to people who "discover things that are surprising—so surprising that those things make people LAUGH, then THINK."[1] Those who win Ig Nobel awards are not widely lauded for their achievements, and their names will probably never be mentioned in history books. But they do get attention in blogs, podcasts, videos, and magazines.

German psychologist Fritz Renner won an Ig Nobel in 2025 for his research that showed people's ability to speak a foreign language improves after they drink a small amount of alcohol. Renner does not encourage people to consume booze to become more fluent, but he does believe in the merit of Ig Nobel Prizes. "It's not all about groundbreaking discoveries, but also about carefully examining everyday assumptions," Renner says. "In a time when misinformation and 'fake news' make it hard to separate opinion from evidence, that feels especially worth celebrating."[2]

Studying Cow Burps

One of the most important scientific issues sullied by misinformation is climate change. And Ermias Kebreab, a professor and director of the World Food Center at the University of California, Davis, is guiding research that makes people laugh before they think. Kebreab's main focus is cow burps. And he helped design a strange-looking contraption to learn more about them.

> **"[Ig Nobel Prizes are] not all about groundbreaking discoveries, but also about carefully examining everyday assumptions. In a time when misinformation and 'fake news' make it hard to separate opinion from evidence, that feels especially worth celebrating."[2]**
>
> **—Fritz Renner, psychologist**

Anyone walking into the barn on the University of California, Davis, campus in 2019 might have chuckled when they saw the neck and head of a Holstein cow sealed in an airtight clear plastic chamber. The cow was calmly eating its feed as large tubes pumped air in and out of the chamber. The strange-looking scene was part of an experiment with world-changing potential. The air exiting the chamber was sent to a nearby lab, where specialized equipment measured the exact amount of methane gas in each cow belch.

The large plastic box around the cow's head helped researchers determine that a cow can burp up 220 pounds (100 kg) of methane in a year. Methane is a potent greenhouse gas that contributes to climate change. The methane produced by US cattle accounts for 4 percent of all greenhouse gas emitted annually in the United States.

Kebreab knew from previous studies that seaweed can inhibit the microbes that produce methane in an animal's digestive tract.

A worker at the University of California–Davis prepares to provide a cow with feed that includes seaweed. Researchers found that adding seaweed to a cow's diet reduced the amount of methane the animal produced by 80 percent.

The University of California, Davis, experiment allowed scientists to test the theory on live cows. Researchers found that a diet that incorporated seaweed reduced the amount of methane a cow produced by 80 percent. As Kebreab says, "Climate change is happening and it's not waiting for anybody. We have the solution. . . . I believe that we can reach into a climate-neutral situation where what we eat . . . does not have an impact on the warming of the climate."[3]

Momentous Discoveries Mocked

The role of cow burps in climate change has been mocked on television talk shows, social media platforms, and even the floor of the US Senate. Those who follow the science are sometimes ridiculed, erroneously, for studying cow farts (when 97 percent of cow methane comes from burps). But throughout history, scientific experiments that were initially viewed as absurd have led to momentous discoveries. In 1847 Hungarian physician Ignaz Semmelweis proposed the idea that deadly infections could be greatly reduced if doctors simply washed their hands before working with patients. This went against common practices at the time. Semmelweis was hounded by the medical profession. He suffered a mental breakdown, dying at age forty-seven in an asylum. Today Semmelweis's discovery is common knowledge—infections can be spread by harmful bacteria on a doctor's hands.

Research has come a long way since the nineteenth century. Today most researchers follow a process called the scientific method. The scientific method begins when scientists pose a question. They develop theories that might answer the question and conduct experiments to test the theories. Scientists analyze the data from the experiments and draw conclusions. If experiments fail, scientists will often return to the problem with new or refined procedures.

Science is a dynamic process. Conclusions reached through the scientific method are often reexamined by other scientists who apply new data or technologies to existing theories. Challenging what seem like settled facts can lead scientists to seek knowledge in strange new places. Processes that might seem bizarre to casual observers today can lead to discoveries that change the world. The Ig Nobel Prizes celebrate that atypical line of thinking.

CHAPTER ONE

Yucky Stuff

Jennifer Macalady loves slime. And she thinks slime is so cool that she decorates her office with bottles of purple, black, and greenish sludge. She says the substances look like smoothies that have been left on the counter for several days—or weeks. Macalady is a scientist whose tongue-twisting job titles include geomicrobiologist, biogeochemist, and astrobiologist. All three specialties involve studying the slimy microorganisms that are the source of all life on earth—and in the case of astrobiology, life in the universe.

Macalady finds the best slime when exploring caves in Italy, Mexico, Bahamas, and the United States. When cave exploring, Macalady descends deep in the bowels of the earth, where the rotten-egg stink of sulfide gas would make most people turn back. But Macalady finds beauty in the slippery mud and the disgusting snot-like strands of goo that hang off cave ceilings. She recalls that the first time she hunted cave slime in 2002, she lost track of time and stayed beneath the earth for eight hours: "I was absolutely smitten, just starry-eyed. There was so much slime of so many different kinds, on every surface. It was a completely transformative experience."[4]

"I was absolutely smitten, just starry-eyed. There was so much slime of so many different kinds, on every surface [of the cave]. It was a completely transformative experience."[4]

—Jennifer Macalady, biologist

Macalady understands that billions of years ago the environment on earth was similar to caves she explores. There was little oxygen, and the main lifeforms were various types of slime. These microorganisms eventually evolved into more complex organisms. And this type of slime evolution is most likely taking place on countless planets throughout the universe.

Only around 1 percent of all microbial species have been identified by biologists, and caves are some of the least explored places on earth. Macalady says her love of slime has led her to find new species. She hopes these slimy microbes lead to a better understanding of the planet's past. "Because we don't see [microbes], we don't have this gut feeling for how much a part of Earth's life they are. . . . There is an enormous amount of potential to learn new and useful things about Earth life by continuing to pick away at [slime],"[5] she says.

Planet-Saving Slime

Some of the slimy microbes living in caves are known as extremophiles. These organisms flourish in extreme environments that might include intense cold or heat, total darkness, or high pressure. And the slime-loving microbiologists James Henriksen and Braden Tierney know they do not need to go cave diving to find extremophiles for research. Instead, they mine for slime in showerheads, dishwashers, water heaters, and deep fryer drip pans.

Henriksen and Tierney are focused on finding unknown microbes that might help slow climate change. They were inspired to start the search after studying microorganisms that grow around volcanic vents under the ocean. These vents constantly expel carbon dioxide, the main gas responsible for global warming. After analyzing ocean waters around the vents off the coast of Italy, Tierney discovered an organism he named Chonkus. It grows fast by eating carbon dioxide. When Chonkus is full it sinks to the bottom of the ocean, keeping the carbon out of the atmosphere.

Tierney and Henriksen were prompted to found the Extremophile Campaign: In Your Home. The campaign's website asks the public to "look in and around your home for any signs of weird microbial growth (think slime, crusty mats, stringy growth). Take a picture of it and answer a few questions about what you see."[6] People are encouraged to search for slime and crust near frigid air conditioners, dank toilet water tanks, hot dishwasher pipes, and moldy basements. If researchers find anything interesting in the photos, they send kits to the contributors, who are instructed to collect and mail samples back for further study.

This purple creature is a type of organism known as an extremophile because it thrives in extreme environments that may include intense cold or heat, total darkness, or intense pressure.

Henriksen says he got the idea to search for extremophiles in homes after reading about a homeowner with a particularly strange problem—slimy tentacles growing out of their showerhead. The subject of the paper, he notes, "had to be horrifying for the person and hilarious for the scientists."[7]

Brian Orr is a contributor to the Extremophile Campaign. Orr is an HVAC specialist, a person who services heating, ventilation, and air-conditioning systems. While working in the hot, humid Florida climate, Orr says he has handled countless pipes clogged with "gelatinous glop . . . [similar to] elephant snot—a clear or white sort of goo that would build up really, really fast."[8] Orr was among the approximately 120 people who contributed to the Extremophile Campaign in 2025. Researchers have sequenced the DNA on over forty-five samples to see whether any might be helpful in slowing climate change. While some might recoil when they see crusty mats and stringy goo, some of the elephant-snot slime under the sink might contain the key to cooling the climate.

Weird and Wonderful Fatberg Photo

The Ig Nobel Prize is characterized by focusing on strange and silly research. But the Ig Nobels are not the only prizes coveted by lovers of weird science. In the United Kingdom the Engineering and Physical Sciences Research Council grants awards in the Weird and Wonderful category. In 2019 that award went to Natalia Jawiarczyk for her unusual photograph of a fatberg sample in a laboratory. As a biotechnologist, Jawiarczyk experiments with methods for preventing fatbergs from forming in sewers. Her photo was of a small portion of the gigantic Whitechapel fatberg that made headlines in 2017. The picture showed fatberg crystals on filter paper with various elements separated out using an organic solvent. Jawiarczyk says a fascinating aspect of her research is

> working with the "beasts from beneath." As part of my research I have had the opportunity to analyze pieces of the famous Whitechapel fatberg. . . . Currently the big unknown is around the reactions leading to fatberg formation, and this is where my research will have impact for the public, the environment and the water companies. . . . Tackling the science behind deposit formation will enable much more effective and targeted sewer . . . management.

Quoted in Aeroemma, "Be like Water: Meet Natalia Jawiarczyk," Soapbox Science, June 11, 2019. https://mksoapboxscience.wordpress.com.

Pretty Smart Slime Mold

It remains an open question whether gelatinous glop can help solve climate problems. But Japanese biologist Toshiyuki Nakagaki has found that slime mold has other unexpected uses. In 2008 Nakagaki and his colleagues won an Ig Nobel Prize in Cognitive Science for showing that brainless blobs of goo have intelligence. And they are clever enough to navigate a maze.

Nakagaki's team focused on a *Physarum polycephalum*, a single-celled organism large enough to be seen by the naked eye. According to Nakagaki, the yellow mold "slimes its way through the world seeking food."[9] Slime mold can be found in nature feeding on the bacteria and fungi that live in rotting logs and among decaying leaves in forests. When slime mold discovers a piece of food, it surrounds it with a network of tubes similar to blood vessels that distribute nutrients to cells.

To study the network of nutrient tubes, researchers placed slime mold in a small maze. The maze was covered in food, which allowed the slime to spread its tubes along every path. In the second part of the experiment, the food was removed everywhere but at the end of the maze. The slime mold was still able to find the shortest way through the maze to reach the food. As Nakagaki points out, "Humans are not the only living things with information-processing abilities. . . . Simple creatures can solve certain kinds of difficult puzzles. . . . If you want to spotlight the essence of life or intelligence, it's easier to use these simple creatures."[10]

"Humans are not the only living things with information-processing abilities. . . . If you want to spotlight the essence of life or intelligence, it's easier to use these simple creatures."[10]

—Toshiyuki Nakagaki, biologist

But Nakagaki was not done with slime mold. In a 2010 experiment, which also won an Ig Nobel Prize, Nakagaki's team used slime mold to design an effective railway network. The organisms seem to love oat flakes, so this type of food was used to create a map of Japanese cities that surround Tokyo. The mold explored the map for several hours, then sent out nutrient-delivery tubes throughout the oat flakes. After a day, some of the unused tubes

Japanese biologist Toshiyuki Nakagaki discovered that slime mold, like the sample shown here, has enough intelligence to navigate a maze.

Pee-Powered Fuel Cells

Human waste pollutes rivers, poisons water supplies, and creates a stench. But in 2017 scientists at the University of the West of England in Bristol found a new use for sewage. They invented a microbial fuel cell that runs on urine. Traditional fuel cells are batteries that generate electricity by splitting hydrogen molecules into electrons and protons. The electrons produce electrical energy. Microbial fuel cells contain a species of bacteria that feeds on the organic material found in urine. As the bacteria feed, they produce an electric current. While the process might sound disgusting, the benefits are real. There are over 2.5 billion people in the world without access to clean toilets and 1.2 billion without electricity. The English scientists installed microbial fuel cells in the bathrooms at a school in Uganda. The cells power the lights in the bathroom. As lead researcher on the project Ioannis Ieropoulos explains, "This technology not only cleans the wastewater, and so improves sanitation and hygiene, but at the same time it is generating energy. . . . There is a huge amount of wastewater that goes down the drains every hour around the world. This is where the greatest potential lies for the technology."

Quoted in Richard Gray, "'Fatbergs,' Faeces and Other Waste We Flush Could Be Fuel," BBC, October 5, 2017. www.bbc.com.

fell away. The interconnected tubes that remained were nearly identical to a map of the intercity Japanese railway system.

A member of the research team, Atsushi Tero, says that studying slime mold might seem silly. But the way the organisms plan food networks might be imitated when creating new roads, railways, or decentralized communications networks. According to Tero, "Slime molds, without calculating all the possible options, can flow over areas in an impromptu manner and gradually find the best routes. Slime molds that have survived for hundreds of millions of years can flexibly adjust themselves to a change of the environment. They can even create networks that are resistant to unexpected stimulus."[11]

The Whitechapel Fatberg

City planners in London, England, might look to slime mold when planning an updated sewer system. The city's sewers, which were built in the mid-nineteenth century, are out-of-date for the modern world of waste. This fact was underscored in 2017 when London sewer workers made a disgusting discovery under Whitechapel

Road. The sewer was blocked by a 143-ton (130 metric ton) blob of repulsive gunk. It was composed of human waste, cooking oil, wet wipes, and other nastiness that makes its way down the city's toilets. Known as the Whitechapel fatberg, the revolting fatty mess was 820 feet (250 m) long.

Fatbergs are so named because they are as hard as icebergs and as difficult to break up. Basic science is behind fatberg formations. Solidified fat, oil, and grease, known as FOG, reacts with calcium, a dissolved mineral found in most water. The combination creates a sludgy compound that binds with other stuff that people flush down the toilet, including cotton balls, wipes, and dental floss. Sewer workers have even found clothing, false teeth, rings, and watches in fatbergs. Sanitation engineers remove the smelly mountains of waste by breaking them into smaller pieces with high-powered water jets, saws, and pickaxes.

> **"We clear a sewer blockage caused by fatbergs every seven-and-a-half minutes. . . . Eight times every hour, a [London water] customer suffers a blockage caused by sewer abuse."[12]**
>
> —Lawrence Gosden, engineer

The Whitechapel fatberg became an internet sensation because of its massive size. But fatbergs are a constant problem for London sewer workers, according to engineer Lawrence Gosden, head of wastewater treatment in the city. "We clear a sewer blockage caused by fatbergs every seven-and-a-half minutes and spend more than [$1.3 million] a month clearing them. . . . Eight times every hour, a [London water] customer suffers a blockage caused by sewer abuse,"[12] he says.

Gosden says workers treat around 366,000 fatberg sewer blockages across the United Kingdom annually. And fatbergs are an increasingly common problem throughout the world. They have caused major sewer problems in New York, Denver, and Baltimore in the US, as well as in Melbourne, Australia and Valencia, Spain, and elsewhere.

Recycling a Fatberg

Whenever disgusting piles of slime cause troubles, scientists can be relied on to provide solutions. It took sewer workers three weeks to remove the Whitechapel fatberg, but they did not dump all of it in a landfill. The human waste bomb was sent to a Scottish company called Argent Energy, where the solid muck was pushed through a gigantic strainer to remove wipes and other solid objects. The remaining 50 tons (45.4 metric tons) of slime was heated to melt the fat and process it into biodiesel. This eco-friendly fuel, which can replace petroleum-based diesel, went on to power London's 350 buses for an entire day.

Scientists at Argent Energy have been creating biodiesel from repulsive waste products since the early 2000s. As Argent development director Dickon Posnett says, "We're always looking for new sources. We've done rancid mayonnaise, old soup stock that has been sitting in tanks for years, and now we're playing with fatbergs."[13]

In addition to creating biofuel, a portion of the stomach-churning Whitechapel fatberg is a museum piece. Since 2018 visitors to the Museum of London have been able to tour an exhibit called *Fatberg!* that includes hunks of dried fatberg in an airtight display case. As the director of the museum, Sharon Ament, says, "It is important for the Museum of London to display genuine curiosities from past and present. . . . This could be one of the most extraordinary objects in any museum collection in London."[14]

Human Waste Moon Base

Turning sewage into biofuel can help slow climate change on earth, but human waste might also be useful for building moon bases. In 2020 researchers at the European Space Agency were examining urine and the role it might play in lunar construction.

There is no water on the moon, and it would be impossible to transport enough water from earth for use in construction. Without water, astronauts cannot construct normal concrete buildings on the lunar surface. But scientists found a possible workaround. Lunar concrete, or lunarcrete, can be made by mixing the moon's powdery soil with urea, the second-most abundant component of human urine after water.

Fatbergs, like the 2017 Whitechapel fatberg, a small part of which is shown here, form when solidified fat, oil, grease, and things that people throw into the toilet react with calcium that is found in most water.

Urea is a popular fertilizer. To test their theory, scientists mixed powdered urea with fine rock particles. The cement substitute, called a geopolymer, was fed into a 3-D printer programmed to create lunarcrete cylinders. The cylinders survived exposure to the extreme temperatures found on the moon, ranging from -414°F to 253°F (-248°C to 123°C). Water remains a necessary ingredient in lunarcrete, but the urea greatly reduces the amount of water needed. According to a European Space Agency statement, "The 1.5 liters (3.2 pints) of liquid waste a person generates each day could become a promising by-product for space exploration."[15]

Studying ways to turn yucky stuff into beneficial products might not be for the squeamish. But when facing the unknown, a puddle of humble sludge might provide a good answer to a puzzling question. No one yet knows whether slime will save the world or urine will build the first moon base. But human curiosity is limitless, and rewards await those who are unafraid to go where no one else has gone before.

Student Guinea Pigs

Guinea pigs are beloved pets. And because they are biologically similar in some ways to humans, guinea pigs are also used for medical research in laboratories. But guinea pigs cannot speak. This makes them useless for psychological experiments. In the 1950s researchers seeking clues about human behavior found that college students made excellent 'guinea pigs.' At the time there were few rules in place to govern how student volunteers could be used to test theories in psychology. In some cases the students experienced little lasting harm. On other occasions students were subjected to bizarre experiments that would be unthinkable in the twenty-first century.

The unusual experiments were conducted when the horrors of World War II were still fresh in the public mind, and there was widespread anxiety about the growing power of the Soviet Union. Researchers driven by these fears set up weird experiments with students and other human guinea pigs, hoping to discover how people could be manipulated to carry out evil deeds.

Following Orders

Faced with the stark horrors of the Holocaust, Yale University professor and social psychologist Stanley Milgram was driven to learn why seemingly average people could participate in mass murder. Social psychologists study how the psychology of individuals can be shaped by cultural practices, social interactions, and society at large. Milgram was a Jewish man born in Brooklyn, New York, in 1933. He was in his early teens when the world first learned of the Nazi's roundup and execution of more than 6 million European Jews from 1941 to 1945.

Like many people, Milgram struggled to understand how average Germans and their collaborators could be made to take part in such atrocities. He understood that the Holocaust could not have happened without the participation of tens of thousands of European citizens. As Milgram wrote, "Gas chambers were built, death camps were guarded, daily quotas of corpses were produced with the same efficiency as the manufacture of appliances. These inhumane policies may have originated in the mind of a single person, but they could only be carried out on a massive scale if a very large number of persons obeyed orders."[16]

A Shocking Outcome

In 1960 Milgram recruited Yale students for a series of experiments that might answer how people could be manipulated to commit atrocities. In the first experiment, students were assigned the role of "teacher" and instructed to address questions to participants called "learners." Milgram, who called himself the "experimenter," was in charge. Teachers were told to press a button that administered an electric shock to learners if the learners answered questions incorrectly. The students were unaware that the learners, who were seated in another room, were paid actors. Learners were told to deliberately answer questions wrong. Students were also unaware that the buttons were harmless props that did not really deliver electric shocks to the learners.

Two of the buttons before the teachers were labeled "Very strong shock" and "Intense shock." Three more buttons, which could supposedly deliver potentially lethal shocks, were labeled "300 volts," "330 volts," and "450 volts." The teacher was told to increase the voltage with each wrong answer. As the (fake) shocks got stronger, the learners pounded on the wall of the adjacent room loud enough for the teachers to hear. When the teachers said they wanted to stop, the experimenter told them to continue.

In a second experiment, the teachers and the learners were in the same room. Learners had to place their hands on a phony shock plate when they answered questions wrong. When they refused, teachers were told to force the learner's hand down to receive the punishment. Milgram described the behavior of one teacher named Bruno Batta, who was instructed to push a distraught learner's hand down onto the shock plate:

> At the 330-volt level, the learner refuses not only to touch the shock plate, but also to provide any answers. Annoyed, Batta turns to him and chastises him: "You better answer and get it over with. We can't stay here all night." . . . The scene is brutal and depressing: his hard, impassive face showing total indifference as he subdues the screaming learner and gives him shocks. . . . When he administers 450 volts, he turns to the experimenter and asks: "Where do we go from here, Professor?" His tone is deferential and expresses his willingness to be a cooperative subject.[17]

When the results of Milgram's experiments were published in 1963, people were outraged. Only five out of forty teachers refused

Social scientist Stanley Milgram performed a series of experiments that tested the willingness of his students to administer what they believed were painful electric shocks to fellow students.

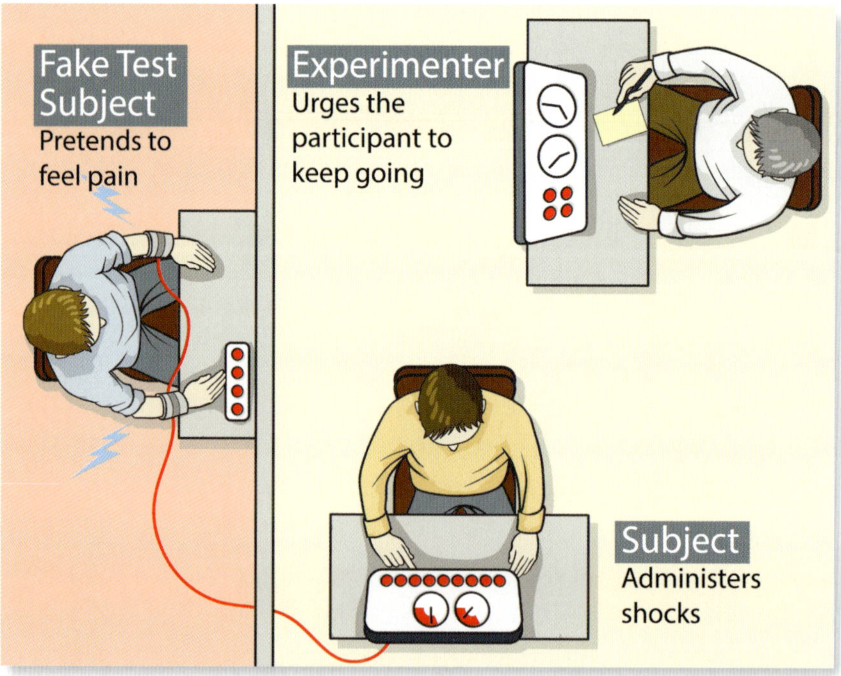

to go to the high shock zones when learners were in a separate room. Fourteen students quit before reaching the 450 volt zone. The other twenty-six students were fully obedient and followed orders by applying the maximum shock to learners. In the second experiment, the obedience rate was 40 percent; sixteen out of forty teachers pressed a learner's hand down on the shock plate.

> **"Naturally poised [undergraduates were reduced to] twitching, stuttering wrecks, on the verge of nervous collapse."[18]**
>
> —Stanley Milgram, social psychologist

Milgram might have run his bizarre experiment with the best of intentions. The investigation seemed to show that most people will follow orders even if obedience to authority caused obvious harm to others. But the Milgram experiment has long been seen as unethical. It should have been obvious that the student participants might suffer long-lasting psychological harm. Milgram could see that happening. He later wrote that the experiment reduced several "naturally poised [undergraduates to] twitching, stuttering wrecks, on the verge of nervous collapse."[18] Test subjects were informed after they participated in the experiment that the shocks were not real, but they were not given further counseling to help them deal with their emotions.

Mystical Experiences

The Milgram experiment ended without social repercussions. The same cannot be said for the experiments conducted by Sidney Gottlieb, a chemist working for the Central Intelligence Agency (CIA). Gottlieb oversaw a program that provided the mind-altering drug LSD to hundreds of unwitting test subjects, many of them college students. Unlike Milgram's experiment, Gottlieb's work had widespread unintended consequences. Some might argue that the course of American culture was permanently changed after LSD was given to a Stanford University student named Ken Kesey.

In 1959 Kesey learned about an unusual experiment being conducted at the nearby Menlo Park Veterans' Hospital. Students were paid seventy-five dollars to take LSD. Lysergic acid diethylamide, known as LSD or acid, was first synthesized by a Swiss chemist named Albert Hoffman in 1938. The drug stimulates areas

Self-Shocking Students

Social psychologist Stanley Milgram is famous for setting up experiments where student volunteers were instructed to give high-voltage electrical shocks to others. The volunteers were unaware that the shocks were not real. But in 2014 students participated in an unusual experiment at Harvard University in which they administered actual shocks—to themselves.

Before the experiment began, students were given a harmless electrical shock and asked whether they would pay money to avoid the sensation. All said yes. Each volunteer was then seated in a chair in an unadorned room. No one was allowed to have books or a cell phone. The students were instructed to do nothing but think for fifteen minutes. The volunteers were told they could break up the monotony by pressing a button that would give them a mild electric shock. Two-thirds of the male volunteers and one-quarter of the women could not last fifteen minutes without pressing the button. The average volunteer pressed it four times, but one student shocked himself 190 times. Psychologist Timothy Wilson, who oversaw the experiment, wrote, "What is striking is that simply being alone with their thoughts was apparently so aversive that it drove many participants to self-administer an electric shock that they had earlier said they would pay to avoid."

Quoted in Ian Sample, "Shocking but True: Students Prefer Jolt of Pain to Being Made to Sit and Think," *The Guardian* (US edition), July 3, 2014. www.theguardian.com.

of the brain that process the senses, including sight, touch, and hearing. Users experience a condition called synesthesia, in which the senses get mixed up. People under the influence of LSD might hear colors, see sounds, and hallucinate intense pulsating patterns and images. Synesthesia also causes what is referred to as the dissolution of the ego. This condition was described by Hoffman in 1943 after he used himself as a guinea pig in a series of LSD experiments: "A portion of the self overflows into the outer world, into objects, which begin to live, to have another, a deeper meaning. . . . [I experienced] the oneness of all living things."[19]

Kesey had never heard of LSD and was ignorant of the drug's powerful effects. Nevertheless, he immediately signed up as a test subject. The experiments were conducted in the psychiatric ward of the Menlo Park hospital. Kesey was led to a small room and given a large dose of LSD. As hidden cameras recorded his reactions, he was interviewed by psychiatrists and others for around eight hours until the drug started to wear off.

Kesey participated in several LSD experiments at the hospital. He had mystical experiences that led him to believe that LSD could help people become more peaceful and spiritually enlightened. But Kesey did not know the source of the drug. In 1953 the CIA began funding what was called Project MK-Ultra, a scheme to conduct a series of drug and mind-control experiments. MK-Ultra was created at the height of the Cold War, when American officials mistakenly believed that the Soviet Union was secretly creating mind-control drugs that could turn average citizens into saboteurs and assassins.

Author Ken Kesey used insights gained while participating in an experiment, in which students were given LSD, to write his best-selling novel, One Flew Over the Cuckoo's Nest.

Students as Prison Guards

One of the most publicized experiments in social psychology is also among the most controversial. The Stanford prison experiment was overseen by social psychologist Philip Zimbardo at Stanford University in 1971. Zimbardo set up the bizarre experiment to study how mentally healthy people can dramatically change when given power over others.

Zimbardo recruited twenty-four student volunteers for the two-week experiment. Some volunteers were assigned the role of prisoners; others were guards in a simulated prison set up in the basement of a campus building. By the second day, most of the guards had lost sight of reality. They began torturing the prisoners, who were stripped naked, chained up, and denied food and sleep. Prisoners began exhibiting signs of anxiety and depression, and some suffered nervous breakdowns. The experiment was called off after six days.

The Stanford prison experiment is repeatedly cited as an example of unethical behavior on the part of Zimbardo, who subjected the students to extreme emotional trauma. Such an experiment would never be allowed today.

A CIA Chemist

MK-Ultra was overseen by Gottlieb, who was seeking a truth serum that would make spies and prisoners of war reveal top-secret information. After Gottlieb tried LSD himself, he used CIA funds to buy up the world's entire existing supply of the drug from Sandoz Pharmaceuticals in Switzerland, where it was manufactured. Throughout the 1950s and early 1960s, Gottlieb oversaw experiments in a specially constructed laboratory, where test subjects were exposed to conditions meant to disorient, confuse, and control them. Gottlieb's first guinea pigs were mostly young prisoners who were given massive doses of LSD, sometimes for weeks or months on end. While under the influence of the drug, they were subjected to interrogations, electric shocks, sensory deprivation, verbal abuse, extreme temperatures, and other forms of torture meant to break their will and take control of their minds. Gottlieb came to understand that LSD could not be used for mind control, but the experiments continued until around 1960.

Gottlieb also used his huge supply of LSD for noncoercive experiments. He set up a phony medical research organization to distribute the drug to clinics and hospitals. Psychiatrists and researchers were told to study the reactions of volunteers on LSD.

The doctors were unaware that the drug came from the top secret MK-Ultra program. And they unexpectedly discovered that LSD could be beneficial to those suffering from anxiety, depression, alcoholism, and other conditions.

Kesey had no idea that he was a guinea pig in a government program overseen by a corrupt CIA chemist. And it did not matter to him. He believed that LSD, which was legal at the time, had the potential to transform society. And he set out to give the drug to as many people as possible.

Kesey knew that the Menlo Park hospital had hundreds of doses of LSD that were easily accessible. He took a job as an assistant at the hospital's psychiatric ward so he could steal the drug. While working at the hospital, Kesey often took LSD while closely observing the patients with mental disorders. His insights led him to write the novel *One Flew Over the Cuckoo's Nest* while working in the early morning hours. The book follows a group of patients emotionally abused by a psychiatric nurse in an evil hospital system. Published in 1962, *One Flew Over the Cuckoo's Nest* was an instant bestseller.

Passing the Acid Test

Kesey used the money from his book to buy a ranch in an ancient redwood forest in La Honda, California, not far from Menlo Park. In 1965 he began his own uncontrolled, bizarre experiments called Acid Tests. These were wild parties held in the San Francisco Bay Area at which attendees were supplied with LSD-laced Kool-Aid. In 1966 Kesey expanded the reach of the Acid Tests, putting on more than a dozen events in auditoriums in San Francisco and elsewhere. Acid Tests featured mind-bending light shows and live music by the Grateful Dead and other psychedelic rock bands. As Grateful Dead lead guitarist Jerry Garcia described it, "It was an open tapestry, a mandala. . . . Thousands of people, man, all helplessly stoned. . . . It was magic, far out beautiful magic."[20]

"[The Acid Tests were] an open tapestry, a mandala. . . . Thousands of people, man, all helplessly stoned. . . . It was magic, far out beautiful magic."[20]

—Jerry Garcia, Grateful Dead guitarist

The Acid Tests helped fuel an international counterculture movement among young hippies. Hippies rejected the traditional values taught them by their parents, schools, and religious institutions. And the social transformation was earth-shattering; several social and political movements, including women's liberation, gay rights, and environmentalism, have roots in the counterculture.

Government authorities saw the counterculture's embrace of LSD as a dangerous threat to society. The federal government

Wild parties known as Acid Tests, which were thrown by Ken Kesey, helped fuel an international counterculture movement among young people known as hippies.

> **"It truly is a . . . curiosity of modern history, that the CIA was responsible for bringing LSD into America, and . . . its effect on the counterculture and on the culture wars that are still continuing today."[22]**
>
> **—Stephen Kinzer, author**

banned LSD use in 1966. But LSD would likely have never escaped the laboratory if not for bizarre experiments conducted on Kesey and other students at the behest of Sidney Gottlieb and the CIA.

Gottlieb died in 1999. His death was noted in obituaries throughout the world. He was referred to variously as a dark sorcerer, the Official Poisoner in Chief, and a mad scientist out of the pages of a cheap spy novel. *The Independent*, a British newspaper, criticized Gottlieb and the CIA: ["Gottlieb] was living vindication for conspiracy theorists that there is nothing, however evil, pointless, or even lunatic, that unaccountable intelligence agencies will not get up to in the pursuit of their secret wars."[21] While intending to defeat a shadowy enemy through mind control, Gottlieb and the CIA accidentally created the first generation of hippies. As Gottlieb biographer Stephen Kinzer wrote in 2021, "It truly is a . . . curiosity of modern history, that the CIA was responsible for bringing LSD into America, and if you unravel everything that LSD meant in the '60s, its effect on the counterculture and on the culture wars that are still continuing today, that's quite a remarkable fact."[22]

Mold, Filth, and Other Medical Miracles

Most people do not love moldy bread, but the ancient Egyptians certainly did. The Egyptians understood that the greenish blue fungus that grows on stale bread has powerful healing properties. Evidence of this belief can be found in the oldest known medical text—the Ebers Papyrus—written in 1550 BCE. The document recommends applying moldy bread to wounds to prevent deadly infections. And many bakers at that time specifically cultivated mold on bread for just that purpose.

The ancient Egyptians probably did not understand that certain molds, such as the genus *Penicillium*, kill bacteria that can cause infections and disease. Today the healing powers of penicillin are well understood, thanks to the work of twentieth-century researchers who specialized in creating mold juice in mold factories. The most well-known mold cultivator, British bacteriologist Alexander Fleming, is credited with discovering penicillin. And like a surprising number of other scientific breakthroughs, the finding was an accident.

In 1928 Fleming took a three-month summer vacation to Scotland. He left behind a messy workbench in London. When Fleming returned, he found that a juicy mold had contaminated a petri dish that he had absentmindedly left on the counter. The dish originally held *Staphylococcus* bacteria, which can cause lethal infections of the skin, lungs, and blood. Rather than toss the mess, Fleming decided to study it under a microscope. He was amazed to see that the green mold prevented the dangerous *Staphylococcus* from growing and multiplying. Fleming called the mold penicillin. He later wrote, "When I woke up just after dawn on September 28, 1928, I certainly didn't plan to revolutionize all medicine by discovering the world's first antibiotic, or bacteria killer. But I guess that was exactly what I did."[23]

> **"When I woke up just after dawn on September 28, 1928, I certainly didn't plan to revolutionize all medicine by discovering the world's first antibiotic [penicillin]."[23]**
>
> **—Alexander Fleming, bacteriologist**

Setting Up a Mold Factory

Despite the groundbreaking discovery, Fleming did not have the resources or the background in chemistry needed to isolate penicillin and develop it as a medicine. Few took notice of Fleming's discovery when he published his research in a medical magazine. Ten years later, in 1938, Howard Florey, a professor of pathology at Oxford University, chanced to read Fleming's research paper while leafing through old medical magazines. He used the information to develop his own strain of penicillin. Florey saw that the drug cured mice infected with sepsis, a life-threatening condition that produces extensive inflammation in the body.

Florey wanted to test penicillin on people. But it takes a large volume of mold culture—528 gallons (2,000 L)—to extract enough penicillin for a single patient. This led Foley and his research team to create a mold juice factory. They cultivated *Penicillium* in bedpans, milk churns, bathtubs, bottles, and empty tin cans. Female workers, referred to as "penicillin girls," were paid to feed and care for the moldy broth. Today penicillin is made in industrial-sized vats in gleaming factories. But according to pharmaceutical executive John L. Smith, it was incredibly difficult to produce large batches of penicillin in the late 1930s: "The mold is as temperamental as an opera singer, the yields are low, the isolation is difficult, the extraction is murder, [and] the purification invites disaster."[24]

Despite Florey's best efforts, his makeshift mold factory could not produce enough penicillin. The situation improved after lab assistant Mary Hunt brought a rotting cantaloupe into the mold factory. Florey said the "pretty, golden mold"[25] on the fruit held a different species of *Penicillium* that was two hundred times

Alexander Fleming is credited with discovering penicillin. Like many scientific breakthroughs, Fleming's discovery was accidental.

stronger than the initial species. The 1939 discovery signaled the birth of the antibiotic age. Several British pharmaceutical companies began manufacturing penicillin, and clinical trials were conducted. In 1942, fourteen years after Fleming's discovery, a woman named Anne Miller became the first person whose life was saved by penicillin.

Before penicillin was produced, there were no effective drugs to treat infections such as blood poisoning, pneumonia, whooping cough, and certain sexually transmitted diseases. Today millions of people are alive thanks to moldy bread, moldy fruit, and dedicated scientists who built a makeshift mold factory.

Feasting on Filth

Those who are made queasy by the thought of mold factories would do well to ignore physician Joseph Goldberger's stomach-churning research methods. Goldberger was an epidemiologist at the US Public Health Service in the early twentieth century. His work focused on preventing the spread of disease among the public.

A malady called pellagra was one of the main diseases that puzzled public health scientists of the period. Between 1900 and 1940, pellagra struck over 3 million Americans, killing more than one hundred thousand. Pellagra is marked by diarrhea, disfiguring rashes, dementia, and death. Because the disease generally afflicted the poor, it was wrongly believed to be caused by unsanitary living conditions. Most doctors mistakenly believed pellagra was spread by infectious germs.

Pellagra was most common in the rural parts of the South. Poor Southerners got most of their calories from three inexpensive food groups known as the three Ms: meal (cornmeal), molasses, and meat (mainly a type of dried pork called fatback). Pellagra was widespread in orphanages and psychiatric hospitals spread throughout Southern states. But the doctors, nurses, and orderlies at these institutions never suffered from pellagra, showing that it was not an infectious disease. This led Goldberger to focus on the three Ms, the basic diet that kept poor people from starving to death. What no one understood at the time is that pellagra is caused by a deficiency of vitamin B_3 (niacin). Most people get enough niacin by eating fresh meat, eggs, grains, vegetables, and dairy products. The three Ms diet provided little niacin.

In 1914 Goldberger set out to prove pellagra was food related. He went to a few orphanages and began feeding the children eggs, milk, peas, and oatmeal. Pellagra cases plummeted. Goldberger published his findings with the recommendation that government agencies provide fresh food to the poor. He assumed this would solve the pellagra epidemic, but he was wrong. Southern politicians were among those who accused Goldberger of attacking their way of life and the economic system that left so many people in poverty.

Goldberger was shocked. To prove that pellagra was not infectious, he organized a bizarre experiment. In 1916 he hosted eight filth parties with around sixteen doctors as guests. The buffet at

> "If anyone ever got pellagra that way, we . . . should certainly have it good and hard! We just feasted on filth."[26]
>
> —Joseph Goldberger, physician

the parties included scabs, blood, and stool and urine samples taken from pellagra victims. The disgusting bodily wastes were injected, swallowed in capsules, or rolled in flour and eaten. Understandably, those who attended the filth parties became nauseous. But after six months no one contracted pellagra. This proved the disease was not infectious and was not spread through unsanitary conditions. As Goldberger later said, "If anyone ever got pellagra that way, we . . . should certainly have it good and hard! We just feasted on filth."[26]

Goldberger's filth parties attracted attention, but critics continued to blame the disease on the purported dirty habits of the victims. This led Goldberger to search for the missing nutrient that caused so much misery. He found that an inexpensive yellow powder called nutritional yeast prevents pellagra. The yeast was used to stop a pellagra epidemic in Mississippi in 1927.

Goldberger died of kidney cancer in 1929. He did not live to see the discovery of the niacin molecule, which was isolated by a biochemist in 1937. Within a few years all bread sold in the

Cured by Sugar Pills

Sometimes patients who are given fake medicine, such as sugar pills, report feeling better. This phenomenon, known as the placebo effect, is attributed to the mind influencing how the body feels. And the placebo effect can occur even when patients know they are taking phony medicine. This was proved by a weird experiment at Harvard Medical School in 2010. Researchers found eighty patients suffering from irritable bowel syndrome (IBS), which causes cramps, diarrhea, bloating, and other intestinal problems. The patients were divided into two groups. One group received no treatment. The other group was told they were taking a regimen of inert sugar pills twice a day. Physician and lead researcher Ted Kaptchuk explains, "We actually had 'placebo' printed on the bottle. We told the patients that they didn't have to even believe in the placebo effect. Just take the pills." After three weeks 65 percent of the patients treated with placebos reported noticeable relief from their symptoms. Some said their improvement equaled that of the most powerful IBS drugs. According to Kaptchuk, "These findings suggest that . . . there may be significant benefit to the very performance of medical ritual."

Quoted in David Cameron, "Placebos Work—Even Without Deception," Harvard Medical School, December 22, 2010. https://hms.harvard.edu.

Eating a balanced diet of grains, vegetables, meat, eggs, and dairy products is key to preventing pellagra, an often-deadly illness that once affected millions of Americans.

United States contained an additive of niacin, and pellagra disappeared. Few people have heard of Goldberg's research, but the pellagra-free populace can thank him for hosting the filth parties that helped eliminate a nasty disease.

The Doctor Who Gave Himself Ulcers

Goldberger was widely ridiculed not just for his filth parties but for questioning uninformed assumptions about a horrible disease. Australian physician Barry Marshall faced similar challenges in 1982 when he gave himself stomach ulcers. Marshall wanted to test his theory about peptic ulcer disease. Peptic ulcers are sores on the lining of the stomach and intestines that cause a gnawing or burning pain, nausea, excessive burping, and bleeding. Peptic ulcer disease is quite common, but until the 1980s doctors wrongly believed sores were caused by excess acid in the stomach. For decades ulcers were blamed on spicy food consumption, stress, and excessive alcohol consumption. Those who had ulcers were given antacids to neutralize the burning pain, but this treatment was largely ineffective.

The Self-Surgery Doctor

Throughout the centuries doctors have used their own bodies to test various medical theories. But Evan O'Neill Kane might go down as having performed one of the most bizarre—and dangerous—self-experiments. Kane was the chief surgeon at Kane Summit Hospital in Darby, Pennsylvania, from the 1880s to the 1930s. During his long career he removed around four thousand infected appendixes from patients. During this era patients were rendered unconscious before an operation with chloroform, ether, or nitrous oxide. These dangerous types of general anesthesia sometimes killed the patients before the operation was concluded. Kane believed he could perform appendectomies using local anesthesia—numbing only the area that would be cut with a scalpel.

When Kane's appendix became infected in 1921, he got a chance to prove his theory by operating on himself. The doctor propped himself up with pillows and injected cocaine into the operating area to numb it. He cut himself open and successfully removed his own infected appendix. But Kane was not done; he seemed to enjoy self-surgery. When he got a hernia in 1932, he once again operated on himself. The seventy-year-old Kane chatted and laughed throughout the procedure. His brother, who was also a surgeon, sewed him up.

Marshall did not think excessive acidity was the cause of ulcers. He conducted research that led him to believe peptic ulcer disease is caused by a corkscrew-shaped bacteria called *Helicobacter pylori*. Animal studies failed to prove his conclusions, so he used himself as a guinea pig. He choked down a dose of *H. pylori* bacteria in a horrible-tasting cloudy broth solution, became extremely ill, and vomited for days.

After more than a week Marshall had an endoscopy, a medical procedure that allows doctors to view the internal organs through a camera fitted to the end of a thin flexible tube. Marshall's theory was correct—the bacteria was multiplying everywhere in his stomach and intestines. As Marshall later wrote, "I had a successful infection. I had proved my point."[27]

A Nobel Prize

In 1983 Marshall published his findings in the British medical journal *The Lancet*. But the medical community considered his self-experiment foolish, and he faced intense criticism for his unorthodox methods. In 1994 Marshall was vindicated for conducting his bizarre experiment. The National Institutes of Health confirmed the link between *H. pylori* bacteria and peptic ulcers. This validation prompted Marshall to write, "I had been waiting for ten years for this day and I felt a combination of relief and satisfaction that I had achieved what I set out to do. Years before, I had developed the hypothesis, tested it, proved it and now it had reached official acceptance."[28]

> **"I had been waiting for ten years for this day and I felt a combination of relief and satisfaction that I had achieved what I set out to do."[28]**
>
> —Barry Marshall, physician

The first antibiotic specifically for use on stomach ulcers was approved in 1995. In 2005 Marshall won the Nobel Prize in Medicine for his discovery.

The Hypervaccinated Man

Most doctors who engage in self-experimentation have a good idea of what they are doing to themselves. But the motives of a man in Magdeburg, Germany, who binged on COVID-19 vaccines remain a mystery. Doctors describe the sixty-two-year-old man as hypervaccinated. He received 217 COVID-19 vaccines over twenty-nine months during 2021 to 2023. That equals about one shot every four days. During that period most people were

Australian physician Barry Marshall revolutionized the treatment of stomach ulcers by proving that these sores, found in the lining of the stomach and intestines, are caused by the corkscrew-shaped bacteria Helicobacter pylori.

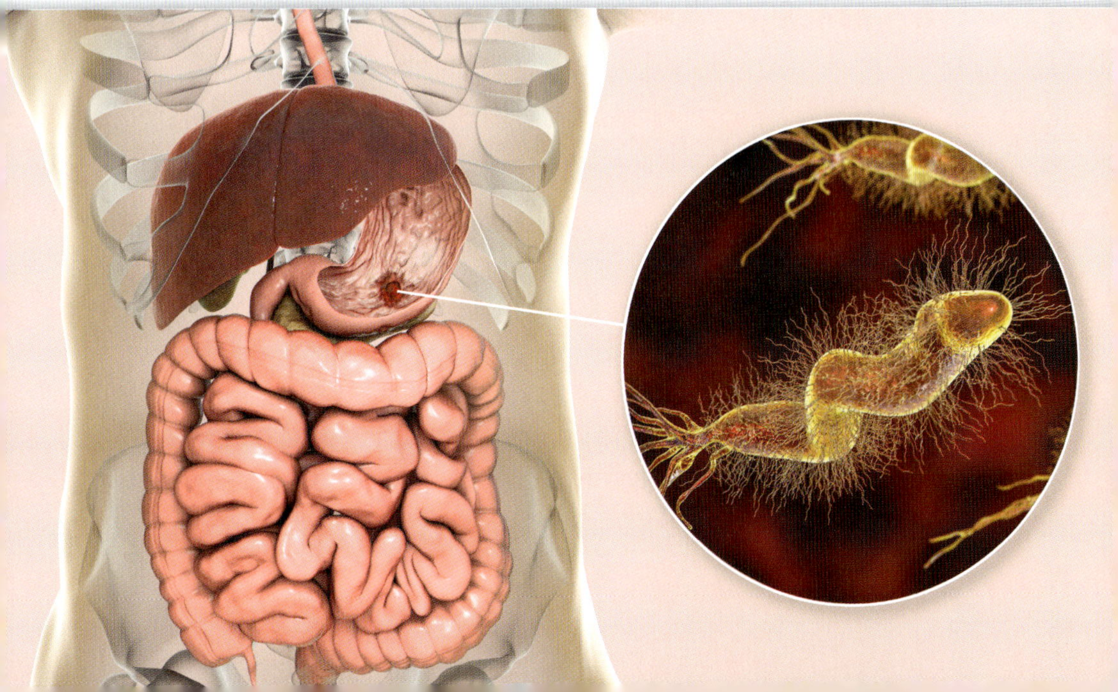

getting one or two COVID-19 shots a year. But the man, whose name has never been released, became a living experiment. He attracted headlines, drew attention from researchers, and ended up under criminal investigation.

During the last six months of 2021, the hypervaccinated man received about one shot per week in Dresden. But in January 2022, he rolled up his sleeves for jabs about once a day, and in the twenty-eight days of February he got thirty-four shots. By this time the man was attracting suspicion. Vaccination centers were issued warnings about the man and were told to turn him away. When the hypervaccinated man began taking his jabs in the town of Eilenberg, he was detained by police. Authorities believed he was selling proof-of-vaccination cards to people who did not want to get the shots. These cards were issued at the height of the COVID-19 epidemic and were required for those who wanted to travel or attend public gatherings. However, criminal charges were never filed against the hypervaccinated man. By May 2023, when the man was 217 shots into his vaccination spree, he agreed to provide blood samples to researchers.

Scientists found that the hypervaccinated man had zero side effects from his self-experiment. And obviously, he never tested positive for COVID-19. The hypervaccinated man proved that the body can easily tolerate COVID-19 vaccines, even in super-high doses. But doctor Kilian Schober says there is no reason anyone would want to follow the man's vaccine routine. He explains, "The benefit is not much bigger if you get vaccinated three times or 200 times."[29]

The hypervaccinated man was lucky—his self-experiment might have had tragic consequences. But his foolhardy research shows that there is a thin line between science and stupidity when it comes to medical self-experiments. Those who succeed might be seen as heroic by some, yet those who fail are forgotten. And anyone contemplating such experiments would be wise to consider the words of physician William Osler: "A physician who treats himself has a fool for a patient."[30]

Investigating Animals

In 2010 livestock researcher Bert Tolkamp was captivated by two fascinating questions. If a cow has been lying down for a long time, is it more likely to stand up soon? And if the cow does stand up, when will it lie down again? In pursuit of scientific answers, Tolkamp assembled a team of cow observation specialists at Scotland's Rural College in Edinburgh.

Tolkamp's team strapped pedometer-like sensors on more than seventy cows. This allowed them to track the movements of pregnant cows housed in barns, dairy cows that were milked every day, and beef cattle that wandered the hills surrounding the campus. The sensors recorded nearly sixty thousand ups-and-downs of cows. Charted actions included lying episodes, standing episodes, lengths of lying, and lengths of standing. To no one's surprise the more time a cow spent lying down, the more probable it was to get back up on its feet. And once the cow was standing, it was very difficult to predict when it would get tired enough to lie down again.

> "By shedding light on cow behavior as an indicator of potential problems, our research is providing new opportunities to protect and improve cattle health and welfare."[31]
>
> —Bert Tolkamp, livestock researcher

The research paper "Are Cows More Likely to Lie Down the Longer They Stand?" won Tolkamp's team the Ig Nobel Probability Prize in 2013. Team members enjoyed winning the annual prize awarded for imaginative research, but the study was conducted for a serious reason. The behavior of a cow reflects its health. Tolkamp explained the importance of studying cows standing, lying, and otherwise moving around: "As farms get bigger, there is less and less direct contact between farm workers and animals, which might mean later detection of problems. By shedding light on cow behavior as an indicator of potential problems, our research is providing new opportunities to protect and improve cattle health and welfare."[31]

Pampered Cow Problems

Cow health and welfare were also the focus of a study released by Japanese researcher Tomoki Kojima in 2019. As a researcher at the Aichi Agricultural Research Center, Kojima was dedicated to finding a way to repel cattle pests. The animals at the center of the study, called Wagyu cows, are raised to be eaten. But their overall happiness is important to those who farm them.

In the United States most beef cattle live miserable lives. They are separated from their mothers at birth and raised on crowded factory farms. Cows are fed hormone-laced feed to pack on weight quickly and slaughtered at a young age. By comparison, Wagyu cows in Japan live lives of luxury. Newborn calves are hand-fed a special milk formula. As they grow, the calves are free to wander with their mothers through rich green pastures. When temperatures fall at night, the calves are dressed in special coats to keep them warm. At seven months the calves are moved to heated barns, where they receive a mixture of rich, natural feed that allows them to grow for around three years. All this pampering promotes small deposits of fat between the muscles, called marbling, which creates buttery, melt-in-your-mouth steaks. And as its price indicates, meat from Wagyu cows is considered to be the most delicious beef in the world. Top grade Wagyu steaks sold for around $200 a pound (454 g) in 2025.

Zebra-Striping Cows

Like all cattle, Wagyu cows attract blood-drinking flies called horseflies that cause pain and transmit disease. In some cases cows become too stressed to reproduce. When a Wagyu farmer contacted Kojima about this problem, a new era in cow research was launched.

In the United States beef cattle are treated with insecticide sprays and powders to prevent fly bites. This is not an option for Wagyu farmers, who refuse to apply toxins to their pampered cows. While searching for solutions to the horsefly problem, Kojima saw a television program about zebras and their stripes. Researchers on the show theorized that the black-and-white markings on zebras helped the animals repel horseflies. For reasons unknown, horseflies are less likely to land on black-and-white surfaces. It is believed that the contrast between dark and light surfaces confuses the flies. This makes it harder for the insects to land.

Kojima wondered whether stripes on a cow would have the same effect. Using nontoxic paint, he painted white stripes on a black Wagyu cow. A second black cow was painted with black stripes, while a third was left unpainted. For thirty minutes twice a day Kojima's team observed the cows and noted their fly-repelling behavior, which included tail swishes, leg stamps, and head shakes. After each observation period, photographs were taken and fly bites were counted.

Researchers found that the white-striped (or zebra-striped) cow received fewer than sixty bites on average, while the other cows were bitten around 110 times. The cow painted with white stripes also exhibited less fly swatting movements. An unnamed official from the local agricultural department commented on the study, "Some producers feel sorry for the cows when they see them being infested with insects and are hesitant to let them graze [in

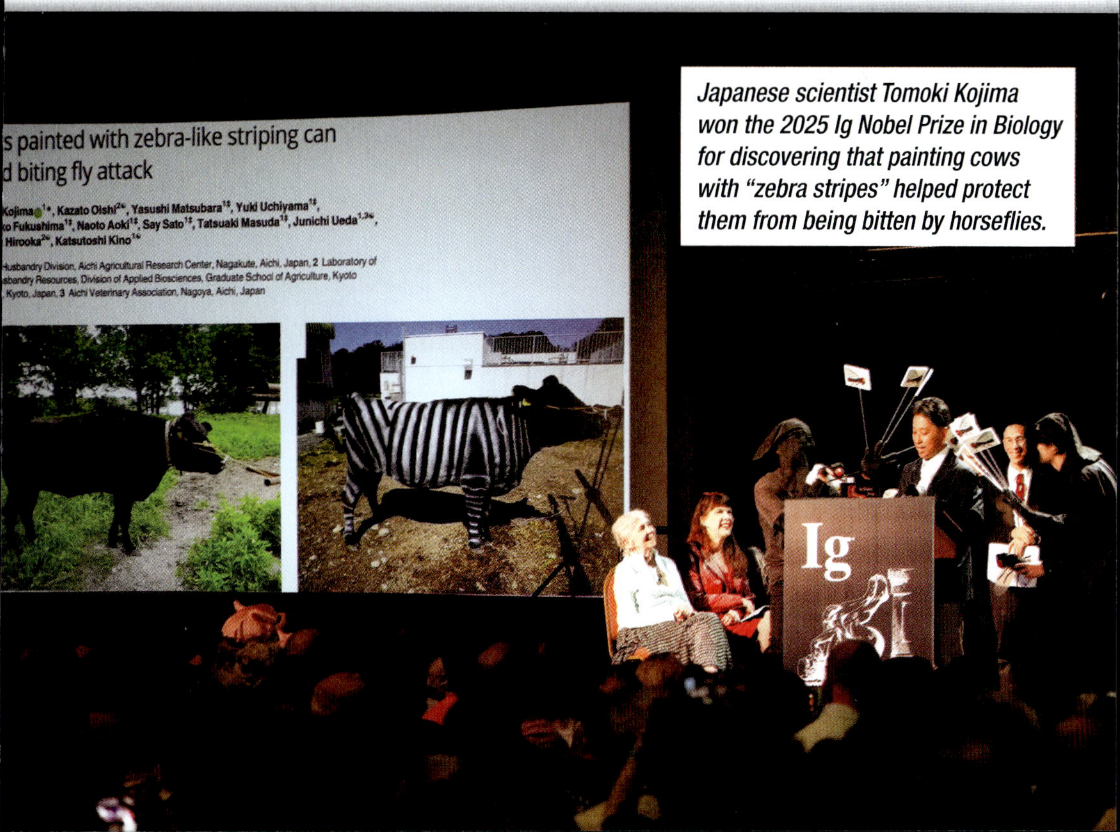

Japanese scientist Tomoki Kojima won the 2025 Ig Nobel Prize in Biology for discovering that painting cows with "zebra stripes" helped protect them from being bitten by horseflies.

"I couldn't believe it. I thought I was dreaming."[33]

—Tomoki Kojima, winner of 2025 Ig Nobel Prize in Biology

open pastures]. By turning them into zebra-patterned 'zebra cows' we can hope that they will be able to eat and rest at their own pace and grow into healthy cattle."[32] In 2025 Kojima won an Ig Nobel Prize in Biology. As he said at the award ceremony, "I couldn't believe it. I thought I was dreaming."[33]

Directional Doggy Doo

Zebra-striped cows can rest more peacefully in grassy fields. And when they do, their behavior will be guided by another kind of field—the earth's magnetic field. Cows are one of several species of animal that possess an ability called magnetoreception. This attribute prompts cows to point north or south when resting or grazing. German researcher Hynek Burda discovered this after analyzing thousands of photographs on Google Earth. The results were the same anywhere in the world where cows could be found grazing. The animals aligned in a north–south direction whatever the weather or time of day.

Deer and mice exhibit the same behavior. Birds, fish, and whales use the earth's magnetic field to navigate. And according to a group of twelve researchers in the Czech Republic, dogs also have magnetoreception abilities. But this is mainly observed when the dogs are peeing and pooping. In 2013 the Czech scientists monitored almost nineteen hundred defecations and fifty-six hundred urinations from seventy dogs of fifty-three different breeds. The scientists learned that both male and female dogs are sensitive to the earth's magnetic field.

Researchers were unable to determine why dogs reject the east–west axis in favor of the north–south when performing bodily functions. But they did not rule out similar behavior in wolves and other canine relatives. The twelve-page report on doggie doo appeared in *Frontiers in Zoology* with numerous charts and graphs. The report concludes, "It is for the first time that magnetic sensitivity was proved in dogs. . . . Our findings open new horizons in magnetoreception research."[34] The deduction, while interesting, might not do much to advance the cause of science. But it does explain why dogs often spin around in circles before excreting bodily waste.

Penguin Poop and Climate Change

Some scientists study how animals poop; others are more interested in what is in the poop. Waste excreted by an animal is referred to as scat or fecal matter, or as guano when produced by seabirds or bats. Whatever it is called, it can tell scientists a lot about a critter's health and well-being. And while analyzing feces might not seem glamorous, it can provide scientists with clues about the ecosystems and environment where the animals live. That is how penguin guano became the focus of Matthew Boyer's studies in 2022.

Boyer is a Finland-based atmospheric scientist who studies climate change. The penguins that attracted his attention live on the frozen Antarctic Peninsula, one of the most isolated places on earth. The species known as Adélie penguins eat fish and tiny shrimplike creatures called krill. With sixty thousand individual penguins living in one area of the peninsula, there was a lot of reeking guano to study.

Atmospheric scientist Matthew Boyer discovered that the ammonia emitted by the guano that builds up around penguin colonies helps keep air temperatures cooler.

Boyer's effort was undertaken specifically to measure the amount of pungent ammonia gas that drifted off the penguin droppings. The ecosystems of Antarctica are threatened by climate change. As temperatures warm the ocean, krill populations are plunging. This results in starvation among penguins and marine mammals. But the ammonia gas that wafts off the penguin guano is slowing the warming where the flightless birds live. Ammonia gas reacts with other chemicals to form what are called aerosol particles. These particles bind with water vapor, which rises into the atmosphere to form clouds. The clouds help keep temperatures lower around the penguin colony. As Boyer explains, "Aerosol particles are necessary for cloud formation; liquid water will not condense to form cloud droplets without the presence of aerosol particles. . . . Clouds influence the . . . surface temperature. Therefore, clouds impact climate change. This is true across the entire planet, not just in Antarctica."[35]

In populated areas, ammonia emitted by farm animals, fertilizer, and industry are a source of danger to human health and the environment. The particles contribute to what is called particulate air pollution, the brown smog that hangs over cities and causes

Getting Bats Drunk

Ethanol is a type of alcohol that naturally occurs when overripe fruit ferments. Birds, insects, and even elephants have been known to get drunk after consuming fermented fruit. This might seem funny, but drunk critters are accident-prone and more vulnerable to predators.

Biologist Francisco Sánchez wondered about the effects of alcohol on wild animals and found that there was little research on the subject. This prompted him to get Egyptian fruit bats drunk. The bats were contained in a large outdoor cage set up with a long flight corridor. The bats were given various amounts of fermented fruit and released. Researchers timed how long it took each bat to fly to the end of the corridor.

Bats use a sense called echolocation to navigate. They send out high-frequency sounds and analyze the echoes bouncing off objects to map their surroundings. Sánchez and his team recorded the echolocation sounds of the boozy bats using ultrasonic microphones. Researchers were not surprised to discover that the more alcohol a bat was given, the longer it took to fly across the cage. And the bat's ability to echolocate was also negatively affected. In 2025 Sánchez and his colleagues were awarded an Ig Nobel Prize in Aviation.

respiratory problems. But there are few sources of human-made particulate pollution in the pristine Antarctic environment. This made the penguin colony an ideal place to study ammonia levels.

Penguins migrate with the seasons. Boyer was surprised to discover that even after the birds moved away from the area, the ammonia levels remained about one hundred times greater than areas where there were no penguins. This provided evidence that penguin poop might be influencing temperatures along a large area of coastal Antarctica while helping reduce the impact of climate change in the region. Boyer explains the importance of his guano-related research: "Our measurements demonstrate that these environmental and ecosystem changes will . . . impact the atmosphere and the climate of the region. This matters, because changes to the climate/ecosystem in Antarctica can have consequences for global climate."[36]

> "I've been a snot collector for almost three years. . . . I basically go fishing for snot. . . . As [beluga whales] come up for air, they exhale, and I collect the sample [from their blowholes]. Then I put the lid on the petri dish, seal it, and keep it in the freezer."[37]
>
> —Justine Hudson, marine biologist

Whale Snot for Science

Many species of whale are also threatened by climate change, and some marine biologists spend their entire careers studying the health of beluga, narwhal, and bowhead whales. One of the slimiest jobs associated with whale well-being falls to professionals who call themselves snot collectors. In this role they collect the mucus contained in the air and seawater that whales shoot from their blowholes. Canadian marine biologist Justine Hudson explained her job in 2020: "I've been a snot collector for almost three years. . . . I made this collection device, a little petri dish attached to a long painter's pole, and I basically go fishing for snot. . . . As [belugas] come up for air, they exhale, and I collect the sample [from their blowholes]. Then I put the lid on the petri dish, seal it, and keep it in the freezer until we fly back home to Winnipeg."[37]

Scientists call whale snot "exhaled breath condensate." The exhaled liquid contains whale DNA, bacteria, and hormones. The

DNA, which is a whale's genetic road map, provides information about the population numbers of various species. Bacteria levels can tell researchers whether whales are sick. And the hormone cortisol gives researchers data about a whale's stress level, which can impact its immune functions, heart, and ability to reproduce.

Hudson's method for collecting whale snot with a long pole is considered old-fashioned. Some marine biologists are employing specially outfitted drones called SnotBots. The first SnotBot was invented in 2021 by Iain Kerr, chief executive officer of the conservation group Ocean Alliance. Kerr got the idea of attaching a petri dish to a drone after returning from a snot-collection expedition smelling of stinky whale slime. SnotBots hover over whales and collect the particles as they blow into the air and rain down on the surrounding waters. Videos from the drones are analyzed to identify individual whales and track their movements, body weight, and overall health. And SnotBots can also be used to track, photograph, and collect whale feces, which reveal information about their diets and toxin levels.

Whales are a keystone species whose well-being is vital to the health of entire marine ecosystems where they live. And collecting

Scientists are analyzing the mucus in the air and seawater that whales shoot from their blowholes. This whale snot contains DNA, bacteria, and hormones that can be studied to gain an understanding of the animals' well-being.

Pizza Lizards

People love pizza. And so do rainbow lizards in the West African nation of Togo, according to research by Daniele Dendi. Rainbow lizards typically dine on insects, but they also tend to like the type of doughy bread used in pizza crust. In 2023 Dendi was at a Togo seaside resort when she saw a rainbow lizard steal a slice of four-cheese pizza from a tourist's table. As Dendi watched the brightly colored reptile gobble down the pizza, she had questions. Do all rainbow lizards like pizza, or was this an isolated incident? Do lizards have favorite toppings? To find out, Dendi acquired nine lizards and let them loose in a large cage. One side of the cage had a tray with a four-cheese pizza. Another pizza topped with ham, mushrooms, olives, and artichoke was set around 30 feet (9 m) away. The lizards loved the cheese pizza. Some were even observed fighting over the last slice. But the lizards shunned the pizza with multiple topping. Researchers determined that the lizards might understand that some foods are easier for them to digest. While no one is sure whether lizards would eat a pineapple pizza, Dendi won an Ig Nobel Prize for her work in 2025.

snot is important work. But is it also fun? As Hudson says, "When we went out for the first season, no one had collected blow from belugas in the wild, so we had no idea if it was going to work. The first day we got a blow sample, we were screaming, 'Yeah we did it!' I have a photo of me holding my first sample, like it was my child. Even if nothing else worked, the fact that we collected that sample was a win."[38]

Hudson's comment reveals the joy experienced by a scientist who will go to great lengths to study a substance most people would rather avoid. Science might not be pretty sometimes, but when the condition of a species or the entire planet needs to be analyzed, scientists go where the information flows—even if it flows from an animal's back end or blowhole.

Frontiers in Food Science

Plastic is everywhere. It is ubiquitous in homes, cars, businesses, and the environment. Plastic is also coursing through the blood and organs of almost every person on earth. Humans constantly ingest tiny pieces called microplastics and nanoplastics that are found in food, food packaging, and the air and water. In 2022 the World Wildlife Fund (WWF) reported that people consume about 5 grams of plastic a week, equal in weight to a credit card.

Since the WWF study was released, countless people have gone out of their way to avoid consuming plastic whenever possible. Some have focused on avoiding a gray plastic substance called Teflon, used to coat nonstick frying pans. Teflon cookware is convenient when making foods that tend to stick to pans. But the chemical used to make Teflon, called polytetrafluoroethylene, is known to have negative health effects.

In 2013 researchers at Rutgers University in New Jersey took a different view of Teflon. They conducted a study to determine whether Teflon might be a beneficial food additive. Physician Rotem Naftalovich led the study. He theorized that people who are trying to stick to a low-calorie diet might feel full faster if Teflon was added to their food. Feeling satiated after a meal is mostly due to the stomach stretching when it is full of food. Adding a zero-calorie additive to food would cause this feeling and help people eat less. Naftalovich wrote that Teflon would be ideal for this purpose because it is "inert, safe, resistant to stomach acid, lacks taste, available in powder form, smooth, resistant to heat, and cost-effective."[39] And since Teflon powder is heat-resistant, it could be baked into a cake or added to a pan of lasagna. And because nothing sticks to Teflon, the material would not bind to the human digestive system. It would harmlessly slide through the body.

To test the plastic-for-dinner theory, lab rats were fed a low-calorie diet that was 25 percent powdered Teflon. After ninety days, the rats seemed to be just fine. There was no sign of toxicity, and the rats lost weight. Naftalovich then used himself as a human guinea pig, eating homemade chocolate-and-Teflon bars. He did not die, and he went on to file for a patent on the idea. Regulators at the US Food and Drug Administration, which oversees food ingredients, were not impressed. Naftalovich later said, "I don't think they wanted to review it because it was such a wonky idea."[40] The "wonky idea" did eventually attract attention, however. In 2025 Naftalovich's research team won the Ig Nobel Prize in Chemistry. In his acceptance speech Naftalovich thanked "esteemed colleagues, confused relatives and curious bystanders. Science is a journey paved with strange questions and sometimes stranger answers. May we never run out of either."[41]

> "Science is a journey paved with strange questions and sometimes stranger answers. May we never run out of either."[41]
>
> —Rotem Naftalovich, physician

Cannibal Calories

The need to feel full after a meal is an age-old problem, as research by British archaeologist James Cole revealed. Cole was vexed by the question of cannibalism among Neanderthals and other early humans. These ancestors to modern people lived forty thousand to four hundred thousand years ago in the Paleolithic era. Anthropologists know that Neanderthals and others were hungry for human flesh. Ancient human bones unearthed in England, Spain, and elsewhere contain marks from stone tools that are very similar to those seen on butchered animals. Some of these bones also have human teeth marks.

Knowing that people once feasted on human muscle and fat, Cole was curious about motivation. Cannibalism today is associated with psychotic criminal behavior. However, Cole wondered whether Paleolithic people butchered others for religious rituals and other spiritual purposes. Or did they revert to cannibalism because they were famished? And if they did eat each other for

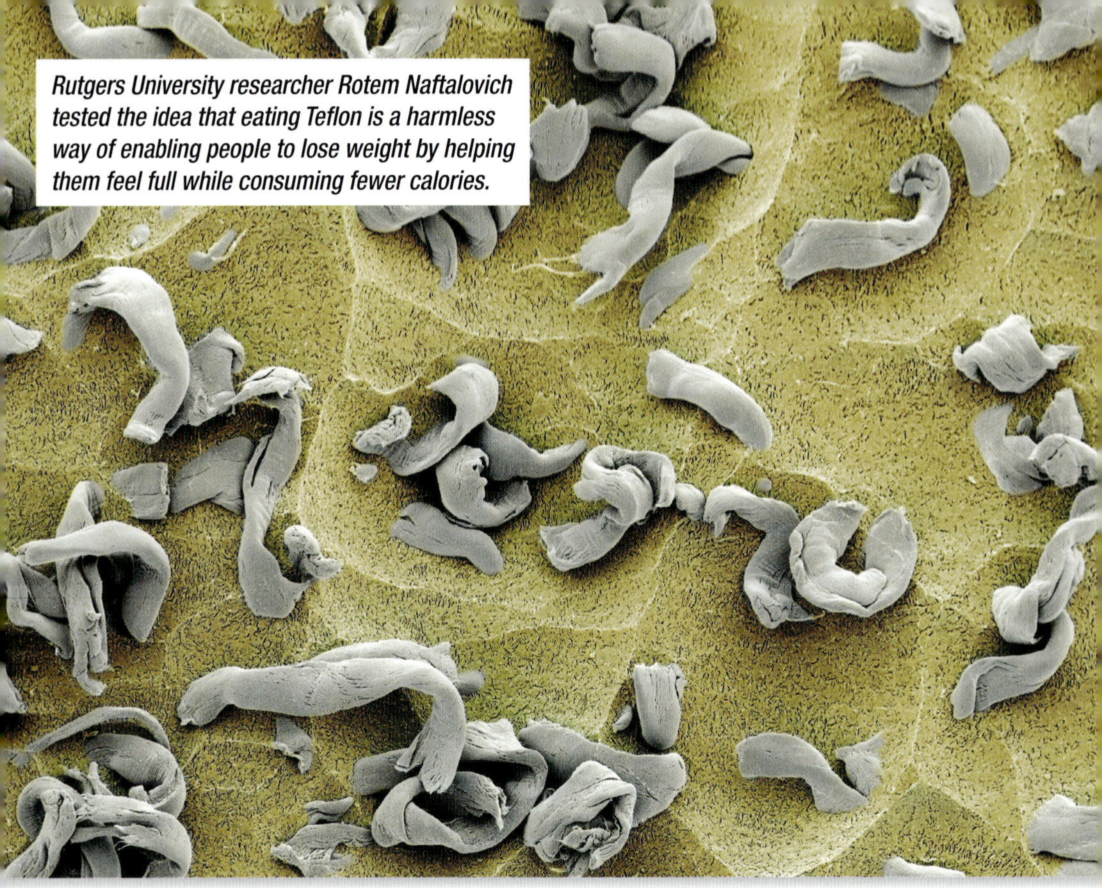

Rutgers University researcher Rotem Naftalovich tested the idea that eating Teflon is a harmless way of enabling people to lose weight by helping them feel full while consuming fewer calories.

nutrition, was human flesh as nutritious as that of other animals? Cole said he developed a morbid fascination with cannibalism in 2007. "I was interested in how nutritious are we actually? Whenever I talk about the topic, I always get a slight sort of side view from my colleagues. . . . [But] if we're calling it 'nutritional' and we have no idea of the nutritional value, then how can that label be used?"[42]

Cole was determined to find out, so he started poring through studies conducted on human cadavers from the 1940s and 1950s. This led him to calculate the caloric content of every part of the body, including fat, muscle, and skin. He determined that the thigh of an average modern male human contains a whopping 13,350 calories. A modern adult male eats around 3,200 calories a day. This means a human thigh could keep a man nutritionally satisfied for

"I was interested in how nutritious are [humans] actually? Whenever I talk about the topic, I always get a slight sort of side view from my colleagues."[42]

—James Cole, archaeologist

more than four days. Cole went on to determine that calves are about 4,490 calories, while the upper arm contains around 7,450 calories. The heart is good for 650 calories, and the liver delivers around 2,570 calories. Kidneys and forearms were also factored into the flesh-eater formula.

Despite the caloric content of a body, Cole reckoned that eating humans strictly for nutritional value was not worth the trouble. A human body could only keep a group of twenty-five men alive for a day and a half. But Paleolithic people hunted and feasted on huge mammals. A single mammoth provided a whopping 3.6 million calories, enough to keep twenty-five people alive for sixty days. A cowlike steppe bison provided ten days of sustenance.

Whether Paleolithic people ate human flesh out of religious beliefs, revenge, or ravenousness remains unknown. Cole believes his calculations on the food value of people are valid, and his efforts won him the Ig Nobel Prize in Nutrition in 2018. He acknowledged that his methods might sound quirky but says the

A Pasta Problem Prize

Biophysicists are experts in math, physics, engineering, biology, and chemistry. They use their knowledge to explore complex issues related to organisms and ecosystems. But sometimes they just want a decent bowl of pasta. One such dish, called *cacio e pepe* in Italian, involves a sauce made from cheese, black pepper, and water. The sauce is supposed to be creamy, but it is challenging to make because the cheese clumps in an unpleasant, stringy mess.

To tackle the pasta problem, biophysicist Giacomo Bartolucci organized a team of highly skilled biophysicists who met on weekends to experiment with cheese, salt, pasta, and hot water. Researchers drew on their knowledge of physics to focus on obscure issues relating to clump formations, starch concentrations, and protein aggregates. As Bartolucci says, "We are all trained theoreticians, so we didn't actually [do] hands-on experiments often. There was no professor involved, and just we wanted to come together and work on this." The experiments led the biophysicists to conclude that adding a small amount of powdered potato starch to the recipe makes the sauce come out nice and creamy. The results, published in *Physics of Fluids* journal, attracted the notice of the Ig Nobel Prize committee, which awarded the pasta scientists the physics prize in 2025.

Quoted in Mary Randolph, "Ig Nobel Prizes Honor Fingernails, Painted Cows and *Cacio e Pepe* in a Celebration of Strange Science," *Smithsonian*, September 25, 2025. www.smithsonianmag.com.

After wondering if Paleolithic people reverted to cannibalism due to starvation, British archaeologist James Cole concluded that was not the case. He calculated that a single mammoth contained 3.6 million calories, enough to keep twenty-five people alive for sixty days.

research was conducted to shine a light on the human condition: "My hope would be that the work would start a discussion around exploring the behavioral complexity of our human ancestors. . . . By doing so, we will be closer to fully understanding the range of behaviors of our ancestors, and in turn get a better understanding of ourselves. This understanding will help us all to build a better and more inclusive future."[43]

Space Miso

Some believe the future is in outer space, but there are problems associated with this vision. People who spend time on the International Space Station (ISS) tend to lose their appetites and stop eating. This might have to do with their diets. Astronauts do not have access to groceries, which are too heavy to shuttle to the space station. And many typical foods spoil without refrigeration. This leaves astronauts to survive on beef jerky, freeze-dried eggs and spinach, and prepared meals such as tuna salad served in pouches. Astronauts spend six months or more living on the ISS, and many would lose their appetite on the diet of a space explorer. Additionally, for unknown reasons, those who spend time in outer space report that their sense of taste gets dull and eating becomes a chore.

Science and soybeans might provide an answer to apathetic astronaut palates. Miso is a traditional Japanese food made from soybeans and is used in soups, sauces, and dressings. Miso is probiotic—it contains microorganisms that are important for gut health. To make miso, chefs ferment salty soybeans for around six months with a fungus called koji. The dark reddish-brown paste that forms is central to Japanese cuisine. It was also the focus of a space-related experiment in 2020. Scientists brewed up some soybeans and koji in small containers and sent them to the ISS. The paste fermented for thirty days before it was shuttled back to earth. Because this was a science experiment, the miso was closely monitored with sensors that detected temperature, humidity, pressure, and radiation levels. Two similar batches of miso were fermented on earth at the same time for comparison.

Microbiologist Joshua Evans from the Technical University of Denmark was one of the scientists involved in the experiment. According to Evans, "We didn't know what to expect—fermentation had never been done before in space. The space miso one was darker and clearly more jostled—which makes sense, as it had traveled much more than either of the earth misos. It was exciting to smell and taste the first bite."[44] The first bite revealed that the space miso was nuttier and had a stronger roasted flavor than the earthbound miso. Evans says this was caused by the radiation of outer space as well as microbes in the koji acting differently in low gravity.

"Fermentation had never been done before in space. The space miso one was darker and clearly more jostled. . . . It was exciting to smell and taste the first bite."[44]

—Joshua Evans, microbiologist

Testing the Five-Second Rule

Most people have heard of the "five-second rule." This hypothetical principle suggests that when food is dropped on the floor, it remains safe to eat if it is picked up within five seconds. In 2003 high school senior Jillian Clark was a summer intern at the food science department at the University of Illinois. She conducted a study to find out whether the five-second rule was true.

Clark purchased gummy bears, fudge cookies, and several floor tiles. She smeared the tiles with harmful E. coli bacteria, which can cause diarrhea, vomiting, fever, and other problems. She dropped the candy and cookies onto the tiles, picked the food up five seconds later, and examined each under a high-powered microscope. The analysis revealed that the all items were contaminated within five seconds or less. Shaking and blowing on the food did not help, since the microscopic bacteria stuck to the candy and cookies. Most kitchen floors are not contaminated with E. coli, but if they are, the five-second rule could lead to intense digestive problems. Clark's study earned her the Ig Nobel Prize in Biology in 2004. She was the first high school student to win the award.

The miso experiment was not the first that focused on creating food on the ISS. Astronauts have grown lettuce and radishes and even threw a taco party after harvesting the first chili peppers grown in space. But brewing miso in space showed that a community of microbes can thrive beyond earth's atmosphere, and that might increase the potential for other forms of life to exist outside the solar system. According to Evans, "By bringing together microbiology, flavor chemistry, [and] sensory science, . . . our study opens up new directions to explore how life changes when it travels to new environments like space. . . . [The research could] enhance astronaut well-being and performance [and] invite new forms of culinary expression."[45]

Dippin' Ice Cream in Nitrogen

When microbiologists start experimenting with foods and flavor, the results can be unexpected, whether in outer space or a garage in Paducah, Kentucky. The garage is where microbiologist Curt Jones discovered a treat that ice cream fans might say changed life on earth.

Jones is the creator of Dippin' Dots, beaded pellets of ice cream that are popular at movie theaters, malls, stadiums, and amusement parks. Most types of ice cream cannot trace their roots to cow feed, but that was Jones's original focus. In 1988 he was in his Paducah garage experimenting with liquid nitrogen, which is the coldest substance that exists on earth. Liquid nitrogen is used in cryogenic technology, which utilizes extremely cold temperatures to change the properties of products. With a temperature of around -350°F (-212°C), liquid nitrogen will flash freeze just about anything. Jones was tossing pellets of grain into the nitrogen in hopes of creating a new way to make animal fodder that would better retain its nutritional value. The experiment was a bust—the cow feed was unusable.

Several weeks later Jones was making homemade ice cream in his kitchen, but he wished he could freeze it faster. Jones picks up the story: "That's when the light bulb came on. I thought, 'I know a way to do that better. I work with liquid nitrogen.'"[46] Jones poured ice cream into a vat of nitrogen. The small ice crystals that formed tasted great and melted in the mouth. Jones patented the creamy pellets under the name Dippin' Dots with the slogan "Ice Cream of the Future."[47]

Most astronauts have to survive on freeze-dried food. However, they have successfully grown a variety of plants, flowers, and vegetables while in space, as shown here in the ISS garden.

By the late 1990s Dippin' Dots were a national sensation. Jones opened around 350 outlets to sell his flash-frozen ice cream. Each location was equipped with a special 40-degree-below-zero freezer that kept the ice cream delicacies at the proper temperature. Jones was bringing in around $20 million annually, but his dreams melted away in 2011. Poor business practices, along with strong competition from a similar product called Mini Melts, drove Dippin' Dots into bankruptcy. A billionaire named Mark Fischer bought the business for $12 million and took it to the next level. Fischer licensed the dot-forming technology to pharmaceutical companies that found ways to extend the shelf life of some medicines through cryogenics. The process was also licensed to cosmetics companies and several plant-based burger brands.

Fischer grew even richer in 2022 when he sold Dippin' Dots to a snack food company for $222 million. Meanwhile, in 2025 Jones was still in the business of providing frozen delights to the public. His company 40° Below Joe sells flash-frozen coffee beads that melt in the mouth and provide a caffeine kick.

Future Foodie Food

Food can be fun and fascinating even as it plays a central role in most people's lives. It is not too surprising that some scientists choose to spend their days experimenting with foodstuffs ranging from ice cream to miso. For those who take an interest in Teflon or cannibals, the work can be less rewarding. But tastes change, and it is impossible to predict what foodies will consider delicious in the future. And if prizes are one day handed out for flash-frozen Neanderthal meat grown in petri dishes in outer space, some scientist with a unique love of the bizarre will likely be there to take credit.

SOURCE NOTES

Introduction: Strange and Surprising Experiments

1. Improbable Research, "About the Igs," 2025. https://improbable.com.
2. Quoted in Chris Simms, "Tipsy Bats and Perfect Pasta Win Ig Nobel Prizes for Weird Science Research," *Scientific American*, September 19, 2025. www.scientificamerican.com.
3. Quoted in Manoush Zomorodi, "What Do Seaweed and Cow Burps Have to Do with Climate Change?," NPR, January 6, 2023. www.npr.org.

Chapter One: Yucky Stuff

4. Quoted in Amanda Paulson, "Cave Diving, Microbes, and Slime: A Love Story," *Christian Science Monitor*, February 20, 2020. www.csmonitor.com.
5. Quoted in Paulson, "Cave Diving, Microbes, and Slime."
6. Sara N, "The Extremophile Campaign: In Your Home," CitSci, 2025. https://citsci.org.
7. Quoted in Ari Daniel, "These Researchers Think the Sludge in Your Home May Help Save the Planet," NPR, June 3, 2025. www.npr.org.
8. Quoted in Daniel, "These Researchers Think the Sludge in Your Home May Help Save the Planet."
9. Toshiyuki Nakagaki et al., "Rules for Biologically Inspired Adaptive Network Design," *Science*, January 22, 2010. www.science.org.
10. Quoted in Shingo Ito, "Japan Scientists Hope Slime Holds Intelligence Key," Phys.org,
11. December 28, 2011. https://phys.org.
12. Quoted in Ito, "Japan Scientists Hope Slime Holds Intelligence Key."
13. Quoted in Ashley Coates, "The Fat-Powered Buses That Are Coming to a Street Near You," *The Independent* (London, UK), August 17, 2017. www.the-independent.com.
14. Quoted in Simon Usborne, "Can We Turn the Whitechapel Fatberg into Biodiesel?," *The Guardian* (US edition), September 17, 2017. www.theguardian.com.
15. Quoted in Patrick Greenfield, "Fatberg Blocking London Sewer Could Become Museum Exhibit," *The Guardian* (US edition), September 13, 2017. www.theguardian.com.
16. Quoted in Nora McGreevy, "Astronauts Could Use Their Own Pee to Build a Moon Base," *Smithsonian*, May 19, 2020. www.smithsonianmag.com.

Chapter Two: Student Guinea Pigs

17. Quoted in Katherine Farmer, "'They Were Just Following Orders,'" UCL, 2025. www.ucl.ac.uk.
18. Thomas Blass, *The Man Who Shocked the World*. Perseus, 2004, p. 184.
19. Quoted in Ben Prawdzik, "Milgram Experiment, 50 Years On," Yale News, September 26, 2011. https://yaledailynews.com.
20. Quoted in Angus Chen, "How LSD Makes Your Brain One with the Universe," NPR, April 13, 2016. www.npr.org.
21. Quoted in Martin A. Lee and Bruce Shlain, *Acid Dreams*. Grove, 1992, pp. 143–44.
22. Quoted in Stephen Kinzer, "Sidney Gottlieb and the CIA Search for Mind Control," History Reader, 2024. www.thehistoryreader.com.
23. Quoted in Ed Prideaux and James Penner, "The Untold Story of the CIA's MKUltra: A Conversation with Stephen Kinzer," *Los Angeles Review of Books*, May 19, 2021. https://lareviewofbooks.org.

Chapter Three: Mold, Filth, and Other Medical Miracles

24. Quoted in Howard Markel, "The Real Story Behind Penicillin," PBS, September 27, 2013. www.pbs.org.
25. Quoted in American Chemical Society, "Discovery and Development of Penicillin," 2025. www.acs.org.
26. Quoted in Markel, "The Real Story Behind Penicillin."
27. Quoted in Sam Kean, "Joseph Goldberger's Filth Parties," *Distillations*, September 8, 2020. www.sciencehistory.org.
28. Barry Marshall, "Barry J. Marshall Biographical," Nobel Prize, 2025. www.nobelprize.org.
29. Marshall, "Barry J. Marshall Biographical."
30. Quoted in Mira Cheng, "A Man Deliberately Got 217 Covid Shots. Here's What Happened," CNN, March 6, 2024. www.cnn.com.
31. Quoted in Johns Hopkins Medicine, "The Founding Physicians," 2025. www.hopkinsmedicine.org.

Chapter Four: Investigating Animals

32. Quoted in BBC News, "Dumfries and Edinburgh Cow 'Pedometer' Study Wins Ig Nobel Prize," September 13, 2013. www.bbc.com.
33. Quoted in Tatsuro Sakata, "'Zebra-Patterned' Bodies Reduce Insect Stress by 70%—Good News for Free-Range Japanese Black Cattle," *Asahi Shimbun* (Nakanoshima, Japan), February 8, 2024. www.asahi.com.
34. Quoted in Andrea Lius, "Zebra Cows Repel Flies and Win Ig Nobel Prize," *The Scientist*, September 18, 2025. www.the-scientist.com.

35. Vlastimil Hart et al., “Dogs Are Sensitive to Small Variations of the Earth’s Magnetic Field,” *Frontiers in Zoology*, 2013. https://link.springer.com.
36. Quoted in Laura Baisas, “How Pungent Poop Could Help Antarctica’s Penguins,” *Popular Science*, May 22, 2025. www.popsci.com.
37. Quoted in Baisas, “How Pungent Poop Could Help Antarctica’s Penguins.”
38. Quoted in Vanessa Minke-Martin, “Coastal Job: Whale Snot Collector,” *Hakai Magazine*, September 20, 2020. https://hakaimagazine.com.
39. Quoted in Minke-Martin, “Coastal Job.”

Chapter Five: Frontiers in Food Science

40. Quoted in Stephen Luntz, “Getting Bats Drunk, Lizards’ Pizza Preferences, and Praising Narcissists Win Big at 2025 Ig Nobel Awards,” IFLScience, September 18, 2025. www.iflscience.com.
41. Quoted in Ian Sample, “Teflon Diet, Garlic Milk and Zebra Cows Triumph at the 2025 Ig Nobel Awards,” *The Guardian* (US edition), September 18, 2025. www.theguardian.com.
42. Quoted in Thomas Sargeant, “Eat Plastic to Lose Weight, Drink Alcohol to Improve Foreign Language Skills: 2025 Ig Nobel Prize Winners,” News.com.au, September 21, 2025. www.news.com.au.
43. Quoted in Nicholas St. Fleur, “Ancient Cannibals Didn’t Eat Just for the Calories, Study Suggests,” *New York Times*, April 6, 2017. www.nytimes.com.
44. Quoted in Kashmira Gander, “Are Humans Nutritious? Cannibalism Study Wins Ig Nobel Prize,” *Newsweek*, September 19, 2018. www.newsweek.com.
45. Quoted in Kathleen Magramo, “Scientists Sent Beans into Orbit and Made ‘Space Miso.’ Here’s How It Tasted,” CNN April 2, 2025. www.cnn.com.
46. Quoted in Magramo, “Scientists Sent Beans into Orbit and Made ‘Space Miso.’”
47. Quoted in K. Annabelle Smith, “Is Dippin’ Dots Still the ‘Ice Cream of the Future’?,” *Smithsonian*, June 10, 2013. www.smithsonianmag.com.
48. Quoted in Smith, “Is Dippin’ Dots Still the ‘Ice Cream of the Future’?”

Books

Melanie Byrd, *Making a Science Geek: A No BS Guide for Teen Girls*. Mango, 2026.

Erika Engelhaupt, *Gory Details: Adventures from the Dark Side of Science*. National Geographic, 2021.

Dan Green, ed., *The Science Book*. DK, 2025.

Nova Kestrel, *Weird Technology: Real Inventions Too Weird to Make Up*. Published by the author, 2025.

Susan Walker, *The University of Bizarre: True Tales of Crazy Research: Laughable Experiments and the Surprising Science Behind Them*. Published by the author, 2025.

Internet Resources

Improbable Research, "About the Igs," 2025. https://improbable.com.

Sam Kean, "Joseph Goldberger's Filth Parties," *Distillations*, September 8, 2020. www.sciencehistory.org.

Vanessa Minke-Martin, "Coastal Job: Whale Snot Collector," *Hakai Magazine*, September 20, 2020. https://hakaimagazine.com.

Sara N, "The Extremophile Campaign: In Your Home," CitSci, 2025. https://citsci.org.

Ed Prideaux and James Penner, "The Untold Story of the CIA's MKUltra: A Conversation with Stephen Kinzer," *Los Angeles Review of Books*, May 19, 2021. https://lareviewofbooks.org.

Manoush Zomorodi, "What Do Seaweed and Cow Burps Have to Do with Climate Change?," NPR, January 6, 2023. www.npr.org.

Websites

BBC Science Focus

www.sciencefocus.com

The *BBC Science Focus* website covers future tech, everyday science, nature, space, and the environment. Strange scientific discoveries are often highlighted, and the *Instant Genius* podcasts feature leading experts who explain ideas in easy-to-understand terms.

Distillations
www.sciencehistory.org/stories/magazine
Unusual scientific stories are the focus of this online publication hosted by the Science History Institute in Philadelphia. Articles cover history, health and medicine, environmental issues, and unusual inventions and discoveries.

IFLScience
www.iflscience.com
IFLScience is dedicated to presenting educational and entertaining science in articles, videos, podcasts, and *Curious* magazine. The site features interesting research on health and medicine, the environment, space, technology, and more.

Improbable Research
https://improbable.com
This website is hosted by the organization that awards Ig Nobel Prizes to those who make surprising and humorous discoveries. The site features podcasts, videos, articles, and publications along with information about every Ig Nobel Prize every awarded since 1991.

Live Science
www.livescience.com
Recent scientific research into space, health, the planet, animals, technology, physics, and more are the focus of this website. The Wild Medical Cases page has numerous articles about weird science and strange discoveries.

The Extremophile Campaign: In Your Home
https://citsci.org/projects/the-extremophile-campaign-in-your-home
This project was founded by microbiologists to encourage people to search for unusual microorganisms that grow in hot, cold, or dark environments in the average home. Anyone can open an account and send slime, crusty mats, and stringy growths to the organization for study.

Note: Boldface page numbers indicate illustrations.

PICTURE CREDITS

Cover: Rich Vintage/Shutterstock

4: Shutterstock
5: Shutterstock
7: Rich Vintage/Shutterstock
11: STEVE GSCHMEISSNER/SCIENCE PHOTO LIBRARY/ Newscom
13: STEVE GSCHMEISSNER/SCIENCE PHOTO LIBRARY/ Newscom
17: PA Images/Alamy Images
20: Steve Lillie/Alamy Images
23: Steve Lillie/Alamy Images
26: Associated Press
30: akg-images/Newscom
33: akg-images/Newscom
35: KATERYNA KON/Science Source
39: Associated Press
41: KATERYNA KON/Science Source
44: Khep/iStock
48: Stefan Diller/Science Source
50: Shutterstock Images
53: Science History Images/Alamy Images

ABOUT THE AUTHOR

Stuart A. Kallen is the author of more than four hundred nonfiction books for children and young adults. He has written on topics ranging from the theory of relativity to making a career in the music industry. In 2018 Kallen won a Green Earth Book Award from the Nature Generation environmental organization. In his spare time, he is a singer, songwriter, and guitarist in San Diego.